Monkeys

Edited by
Lynn Hughes

Workman Publishing
New York

Reprinted by arrangement with
Workman Publishing Co., Inc.

Library of Congress Cataloging in Publication Data

Main entry under title:

Monkeys.
1. Monkeys—Quotations, maxims, etc. 2. Monkeys in art.
I. Hughes, Lynn.
PN6084.M57M6 080 80-13890
ISBN 0-89480-098-1

Book design: Nick Thirkell
Picture research: Ian Wyles

Workman Publishing
1 West 39 Street
New York, New York 10018

Manufactured in the United States of America
First printing September 1980
10 9 8 7 6 5 4 3 2 1

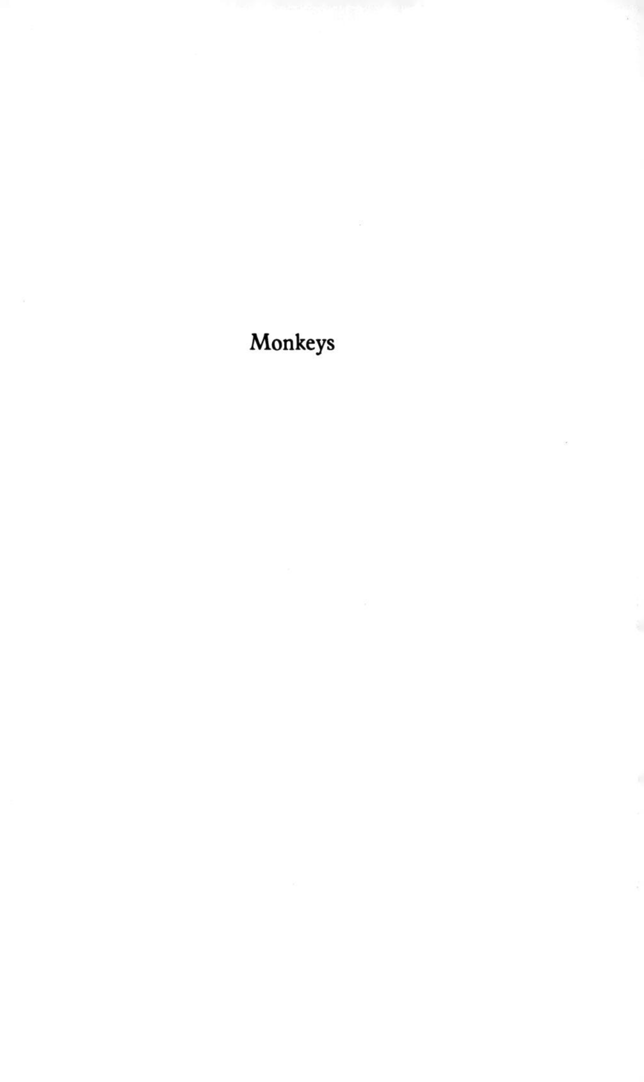

Monkeys

O lovely O most charming pug
Thy gracefull air and heavenly mug
The beauties of his mind do shine
And every bit is shaped so fine
Your very tail is most devine
Your teeth is whiter than the snow
You are a great buck and a bow
Your eyes are of so fine a shape
More like a christains than an ape
His cheeks is like the roses blume
Your hair is like the ravens plume
His noses cast is of the roman
He is very pretty weoman
I could not get a rhyme for roman
And was oblidged to call it weoman.

MARJORY FLEMING (Aged 8)
TO A MONKEY

For the king had at sea a navy of Tharshish with the Navy of Hiram: once in three years came the navy of Tharshish, bringing gold and silver, ivory, and apes, and peacocks. So king Solomon exceeded all the kings of the earth for riches and for wisdom.

I KINGS X 23
HOLY BIBLE

The species Man and Marmozet
Are intimately linked;
The Marmozet survives as yet,
But Men are all extinct.

HILAIRE BELLOC
THE MARMOZET

As a young student, I kept, in my parents' flat in Vienna, a magnificent specimen of a female capuchin monkey named Gloria. . . . One evening, when I returned home after a longer absence and turned the knob of the light switch, all remained dark as before. But Gloria's giggle, issuing not from the cage but from the curtain rod, left no doubt as to the cause and origin of the light defect. When I returned with a lighted candle, I encountered the following scene: Gloria had removed the heavy bronze bedside lamp from its stand, dragged it straight across the room (unhappily without pulling the plug out of the wall), heaved it up on to the highest of my aquaria, and, as with a battering ram, bashed in the glass lid so that the lamp sank in the water. Hence the short circuit! Next, or perhaps earlier, Gloria had unlocked my bookcase—an amazing achievement considering the minute size of the key—removed volumes 2 and 4 of Strumpel's textbook of medicine and carried them to the aquarium stand where she tore then to shreds and stuffed them into the tank. On the floor lay the empty book covers. . . . In the tank sat sad sea-anemones, their tentacles full of paper.

KONRAD Z. LORENZ
KING SOLOMON'S RING

Well I never, did you ever,
See a monkey dressed in leather?
Leather eyes, leather nose,
Leather breeches to his toes.

ANON

MONKEY

Again, the king of Sambacia told my relative, Samuel Blomartius, that these Satyrs, especially the males, in the island of Borneo, have such daring and such strong muscles, that they charged against armed men more than once, and also against defenceless women and girls. Sometimes they were so fired with the desire of them, that they seized and ravished them over and over again. They are extremely addicted to lust (a circumstance which is common to them and the Satyrs of the ancients). Nay, sometimes they are so wanton and lustful, that the women of India have a greater dread of groves and thickets, than they have of dogs or snakes, for in the former these wanton creatures are always lying hid.

ISAAC SCHOOKIUS
UN-NATURAL HISTORY (1669)

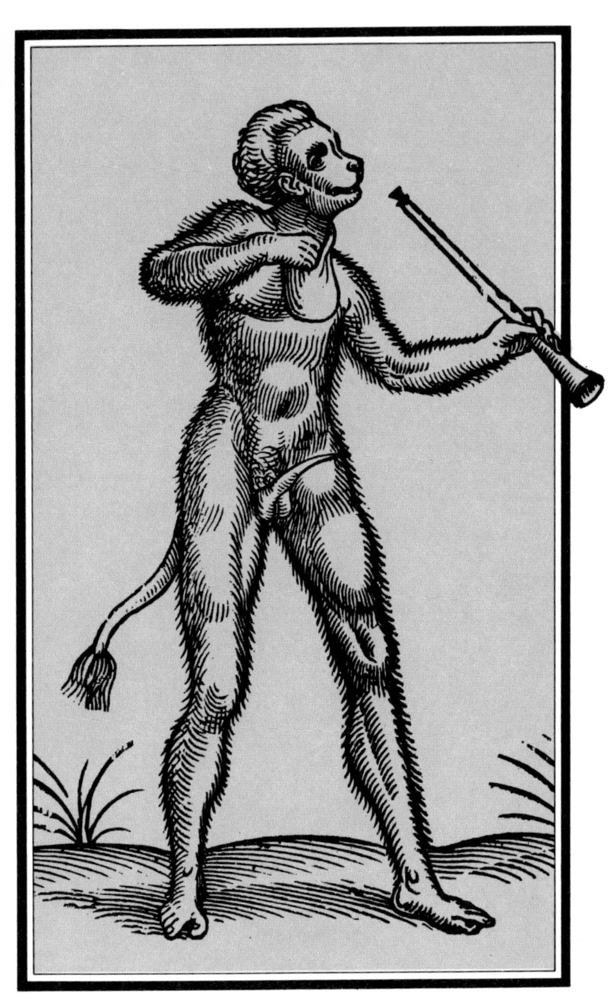

Willie had a purple monkey climbing on a yellow stick,
And when he sucked the paint all off, it made him deathly sick.

MAX ADELER
THE PURPLE MONKEY

217

Monkeys . . . very sensibly refrain from speech, lest they should be set to earn their livings.

KENNETH GRAHAME
THE MAGIC RING

The Big Baboon is found upon
  The plains of Cariboo:
He goes about with nothing on
  (A shocking thing to do).

But if he dressed respectably
  And let his whiskers grow,
How like this Big Baboon would be
  To Mister So-and-so!

HILAIRE BELLOC
THE BIG BABOON

Ah! Sir, you never saw the Ganges—
There dwell the nations called Quidnunkies
(So Monomotapa calls monkeys);
On either bank, from bough to bough
They meet and chat (as we may now);
Whispers go round, they grin, they shrug,
They bow, they snarl, they scratch, they hug,
And just as chance or whim provokes them,
They either bite their friends or stroke them.

JOHN GAY
FABLES

I am a kind of farthing dip
Unfriendly to the nose and eyes
A blue-behinded ape, I skip
Upon the trees of Paradise

ROBERT LOUIS STEVENSON
A PORTRAIT

The constitution of the ape is hot, and since he is rather similar to man, he always observes him in order to imitate his actions. He also shares the habits of beasts, but both these aspects of his nature are deficient, so that his behaviour is neither completely human nor completely animal; he is therefore unstable. Sometimes when he observes a bird in flight, he raises himself and leaps and tries to fly, but since he cannot accomplish his desire he immediately becomes enraged.

SAINT HILDEGARDE OF BINGEN
PHYSICA

Her next victim was a handsome sixteen-year-old girl who had brought a calabash full of snails. The girl, however, was almost as quick in her reactions as Georgina. She saw the baboon out of the corner of her eye, just as Georgina made her leap. The girl sprang away with a squeak of fear and Georgina, instead of getting a grip on her legs, merely managed to fasten on to the trailing corner of her sarong. The baboon gave a sharp tug and the sarong came away in her hairy paw, leaving the unfortunate damsel as naked as the day she was born. Georgina, screaming with excitement, immediately put the sarong over her head like a shawl and sat chattering happily to herself, while the poor girl, overcome with embarrassment, backed into the hibiscus hedge endeavouring to cover all the vital portions of her anatomy with her hands. Bob, who happened to witness this incident with me, needed no encouragement whatever to volunteer to go down into the compound, retrieve the sarong and return it to the damsel.

GERALD DURRELL.
A ZOO IN MY LUGGAGE

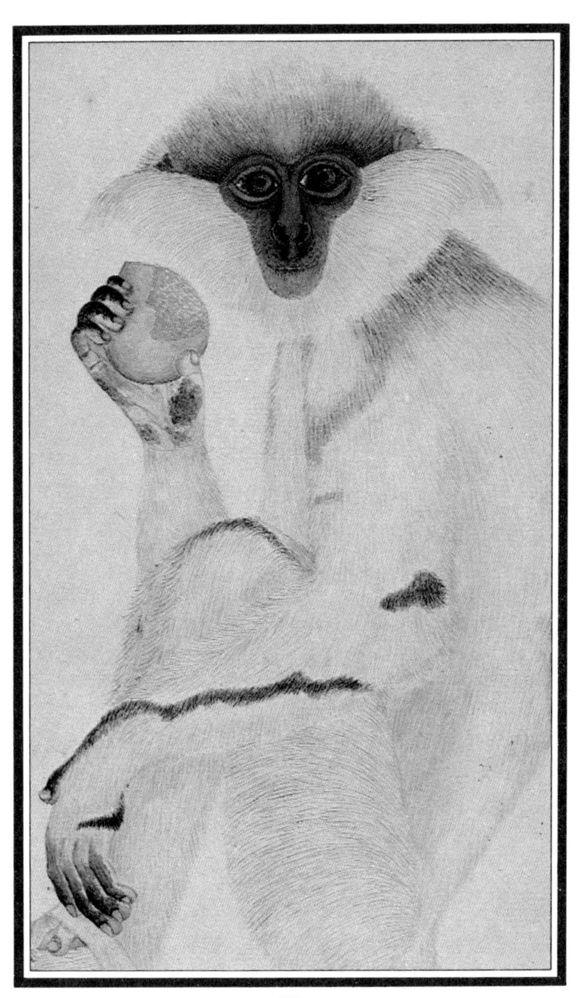

The higher an ape mounts, the more he shows his breech

THOMAS FULLER
GNOMOLOGIA

Healthy wild animals, monkeys are much too clean and active to harbour fleas. When they search one another's fur in a fashion which must be familiar to most persons, they are cleaning their coats of particles of scurf or of similar scraps of dirt and not of fleas. So, speaking generally, it may be said that no fleas have been found truly parasitic on monkeys.

HAROLD RUSSELL
THE FLEA

If someone admired the beauty of the ape, you would say: this is not beauty. And if he were much preoccupied with the way the body of that animal were put together, and admired all its parts as harmonious and fitting, you who know a different degree of beauty would deny this and say; it is not so.

SAINT AUGUSTINE
ENARRATIONES IN PSALMOS

They are called Monkeys (*Simia*) in the latin language because people notice a great *similitude* to human reason in them. Wise in the lore of the elements, these creatures grow merry at the time of the new moon. At half and full moon they are depressed. Such is the nature of a monkey that, when she gives birth to twins, she esteems one of them highly but scorns the other. Hence, if it ever happens that she gets chased by a sportsman she claps the one she likes in her arms in front of her, and carries the one she detests with its arms round her neck, pickaback. But for this very reason, when she is exhausted by running on her hind legs, she has to throw away the one she loves, and carries the one she hates, willy-nilly.

ANON

THE BOOK OF BEASTS (12TH CENTURY)

Black Monky
Mysore Coll. of Drawings

An ape's an ape,
An a varlet's a varlet,
Though they be clad in silk or scarlet.

PROVERB

The Marquess di Bec—ca when appointed to attend the Court of Louis XVI. carried with him a favourite monkey. On his arrival at Paris a great ball was to be given at Versailles to which he was invited. Anxious to excell among the noble guests, he summoned the most celebrated master in the French capital, and applied himself with diligence to attain, under his tuition, a perfect knowledge of the *minuet de la cour*. Pug observed every motion of his master, and chattered at his powered instructor playing on the *kit* while accompanying his pupil through the mazes of the dance. After the lesson was finished the marquess placed a large mirror on the floor and practised before it until he was tired. Pug descended from his observatory, as soon as his master had left the room, put his *chapeau bras* on his own head, took the kit which the professor had accidently left behind him, placed his august person before the mirror, and began to combine the dancing of his master with the chattering and fiddling of his preceptor. Strange discordant sounds being heard to come from the room, the marquess gently approached it, and, through a crevice of the door, detected poor Pug in thus gratifying his vanity by following the example of the embassador.

ANON
MONKEYANA

EXOTICKS
Publish'd according to Act of Parliament May 7, 1761.

The Dog-faced Baboon is from the Coast of Guinea; it is very ferocious and disgusting when fully grown. It has a black visage and long tail. A large specimen died in 1828 in the Tower after attracting much attention . . . from his extraordinary resemblance to humanity, not only in form and appearance, but also in his manners and habits . . . He would brandish his pot of porter, and drink it off apparently with a human relish. His attentions to a dog, that used to be a frequent visitor at his cage, were in the best style of dignified patronage . . . This Baboon, an excessive drinker, at last sunk under a confirmed dropsy.

POPULAR ZOOLOGY (1832)

Chimpanzees generally remain with their mothers till they are two years old, but I have seen three-year-olds still with them. These apes no longer insist on being carried, but at the least alarm they dash to their mothers for protection. Another birth makes the break definite; from then on the mother loses interest in the elder child and devotes her attention exclusively to the younger.

It occasionally happens that young apes only eighteen months old are chased away by their mothers: for most of them this is a death sentence, as, in contrast to elephants who willingly adopt orphans, mother chimpanzees refuse to have anything to do with rejected children.

HEINRICH OBERJOHANN
MY FRIEND THE CHIMPANZEE

At home made my wife get herself presently ready, and so carried her by coach to the fayre, and showed her the monkeys dancing on the ropes, which was strange, but such dirty sport that I was not pleased with it.

SAMUEL PEPYS
DIARY

Women in state affairs are like monkeys in glass-shops.

PROVERB

A Monkey to reform the times,
Resolved to visit foreign climes;
For men in distant regions roam
To bring politer manners home.
So forth he fares, all toil defies;
Misfortunes serve to make us wise.
At length the treacherous snare was laid
Poor Pug was caught; to town convey'd;
There sold. (How envied was his doom,
Made captive in a lady's room!)

JOHN GAY
FABLES

神武天皇即位紀元二千五百五十六年
大 一 三 五 七 八 十 十二
小 二 四 六 九 十一

From the war of nature, from famine and death the most exalted object which we are capable of conceiving, namely, the production of higher mammals, directly follows.

CHARLES DARWIN
ORIGIN OF SPECIES

Illustration acknowledgements

5 *India Office Library and Records.*

7 Melchior D'Hondecoeter: Peacocks. *The Metropolitan Museum of Art. Gift of Samuel H. Kress, 1927.*

9 Nineteenth-century German lithograph. *Andrew Edmunds Collection.*

11 Sobun: Monkey and Octopus. *Victoria and Albert Museum. Crown copyright.*

13 James Marsh: Well I Never. *James Marsh copyright.*

15 The Satyr: From Edward Topsell's *History of Four-footed Beasts.*

17 George Edwards: Black Maucauco. *Trustees of The British Museum (Natural History).*

19 J. B. S. Chardin: The Monkey Painter. *Giraudon/Louvre.*

21 George Shaw: Long-armed Gibbon. *Trustees of The British Museum (Natural History).*

23 Cercopithecus Pluto. *Andrew Edmunds Collection.*

25 Nineteenth-century lithograph of Mandrills. *Andrew Edmunds Collection.*

27 J. B. Audebert: Le Jocko. *Victorian and Albert Museum. Crown copyright.*

29 *India Office Library and Records.*

31 *India Office Library and Records.*

33 Mori-Sosen: Japanese Kakemono. *Shirley Day Ltd.*

35 Eighteenth-century English Print. *Andrew Edmunds Collection.*

37 *India Office Library and Records.*

39 Kunisada: Nineteenth-century Japanese print. *Kokoro, Brighton.*

41 William Hogarth: Exoticks. *Andrew Edmunds Collection.*

43 George Shaw: Variegated Baboon. *Trustees of the British Museum (Natural History).*

45 Japanese Kakemono. *Beasley Collection.*

47 William Hogarth: Detail from Chairing the Member. *Sir John Soane's Museum.*

49 Teniers the Younger: The Sense of Smell. *Private Collection.*

51 Japanese print. The Critic. *Joseph Martin/Scala.*

53 George Stubbs: Rhesus Monkey. *Walker Art Gallery, Liverpool.*

Border line drawings by John Gorham.

For permission to use copyright material we are indebted to the following:

Gerald Duckworth and Co., Ltd. (for the Hilaire Belloc Estate), for 'The Marmozet' and 'The Big Baboon', from *Complete Verse;* Methuen and Co., Ltd., for an extract from *King Solomon's Ring*, by Konrad Lorenz, translated by Marjorie Latzke; Jonathan Cape Ltd., for an extract from *The Book of Beasts*, translated by T. H. White; Robert Hale Ltd., for an extract from *My Friend the Chimpanzee*, by Heinrich Oberjohann, translated by Monica Brooksbank; Hart-Davis MacGibbon Ltd., for an extract from *A Zoo in my Luggage*, by Gerald Durrell.